LEMONLearn
Learners Today. Leaders Tomorrow.

1 2 3 4 5

TODDLER LEAF MATH

WORKBOOKS 19, 20, 21

6 7 8 9 10

LEMON Learn

Learners Today. Leaders Tomorrow.

HOW TO USE THIS BOOK

Encourage your early learner to complete each booklet at their own pace.

1 OPENING PAGE

Each booklet has an opening page. Before starting each section, this page will show you what booklet and skill your child will be focusing on.

2 STEPS 1, 2, 3

Steps 1, 2, 3 are on each opening page to guide you while working with your child.

3 NAME, DATE, TIME

Instruct your learner to write their name, date, and start time before beginning their work. When their work is complete, have them write their finish time. **If the student can not write yet, write it for them.**

HOW TO USE THIS BOOK

Get your student talking! Ask them about the images they see and connections between the workbook and their lives.

4 DIRECTIONS

Have students read the directions aloud on each page, if possible. **If the student can not read yet, read the directions aloud to them.**

LEGEND 5

- Page Number
- Subject
- Booklet Number
- Skill

HOW TO USE THIS BOOK

Students can practice the skills they learn with the color by number pages!

6 COLOR BY NUMBERS

At the end of **Booklets 19 - 21**, you will find a color by numbers page that tests your learner in a fun way.
This activity requires crayons.

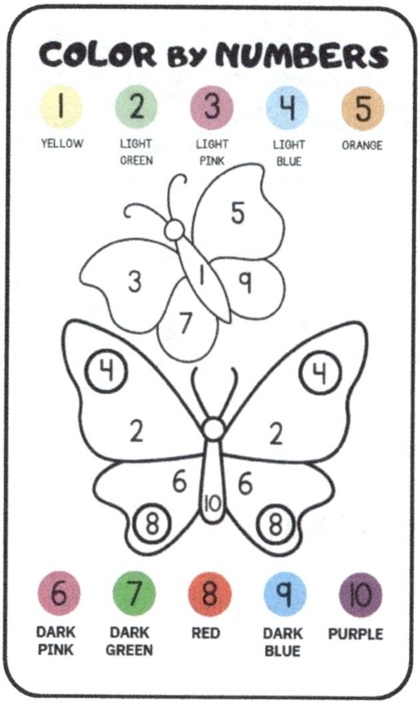

ACHIEVEMENT AWARD 7

At the end of **Booklet 21**, you will find a certificate for your learner. Cut it out and write their name on it. Encourage them to show their friends or hang it on the wall.

CONCEPTS

Booklets 19 - 21

Number Recognition Part 2

- Numerical digits: Recognizing written numerals.
- Writing numbers: Writing numerals by hand.

Fine Motor Skills

Tracing helps children improve their hand-eye coordination, muscle control, and grip strength.

> Learners should work as independently as possible. Guide them to approach each page with a try-again mindset. Every challenge provides an opportunity to learn and grow to achieve success.

FLEXIBILE LEARNING

At Lemon Learn, we understand that every child develops at their own pace, and it's never too early to introduce math—even to children who aren't speaking yet. They can begin this workbook by simply listening to parents recite the oral Booklets 19 - 21 and enjoying the coloring pages that follow. Once your child starts speaking, you can begin again from Booklet 19 to reinforce their learning from the start.

Booklet 19

Tracing up to 10

1 Preview the pages of Booklet 19 with your student before starting.

2 Help your student write their name, date, and start / finish time.

3 Work together with your student for 15-30 minutes daily.

Name: ______________________

Date: ______________________

Time: ____ : ____ **to** ____ : ____

LEMONLearn

Trace the number.

ONE

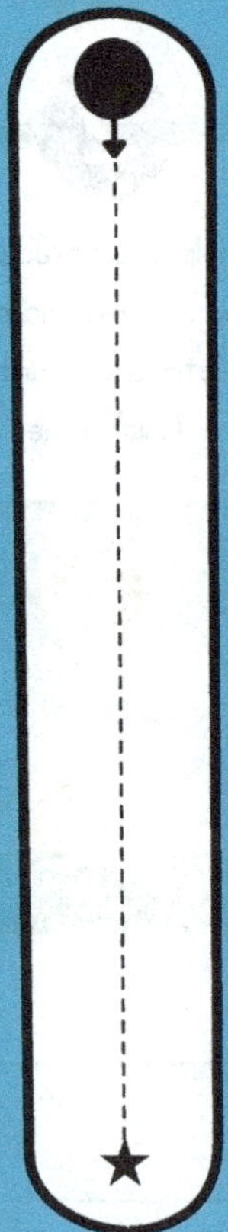

LEMONLearn

Trace the numbers.

ONE

Trace the number.

TWO

182b

LEMONLearn

Leaf Math
Booklet 19
Tracing 1
(up to 10)

Trace the numbers.

TWO

Trace the number.

THREE

Trace the numbers.

THREE

Trace the number.

FOUR

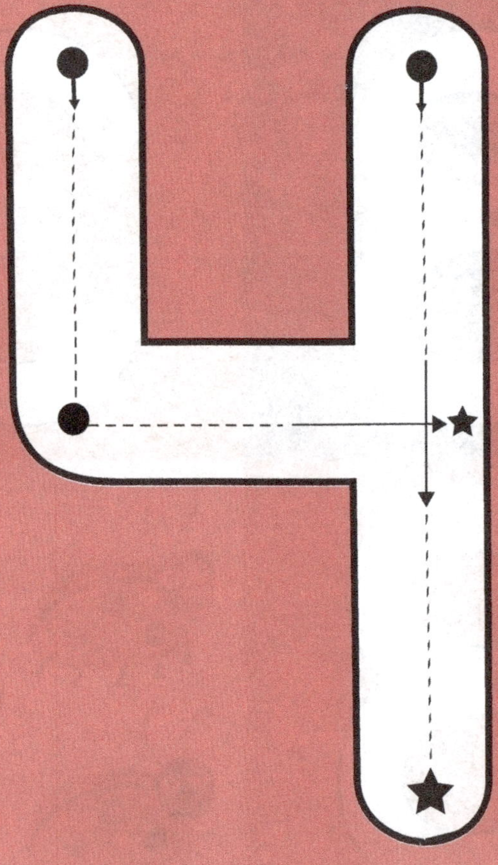

Trace the numbers.

FOUR

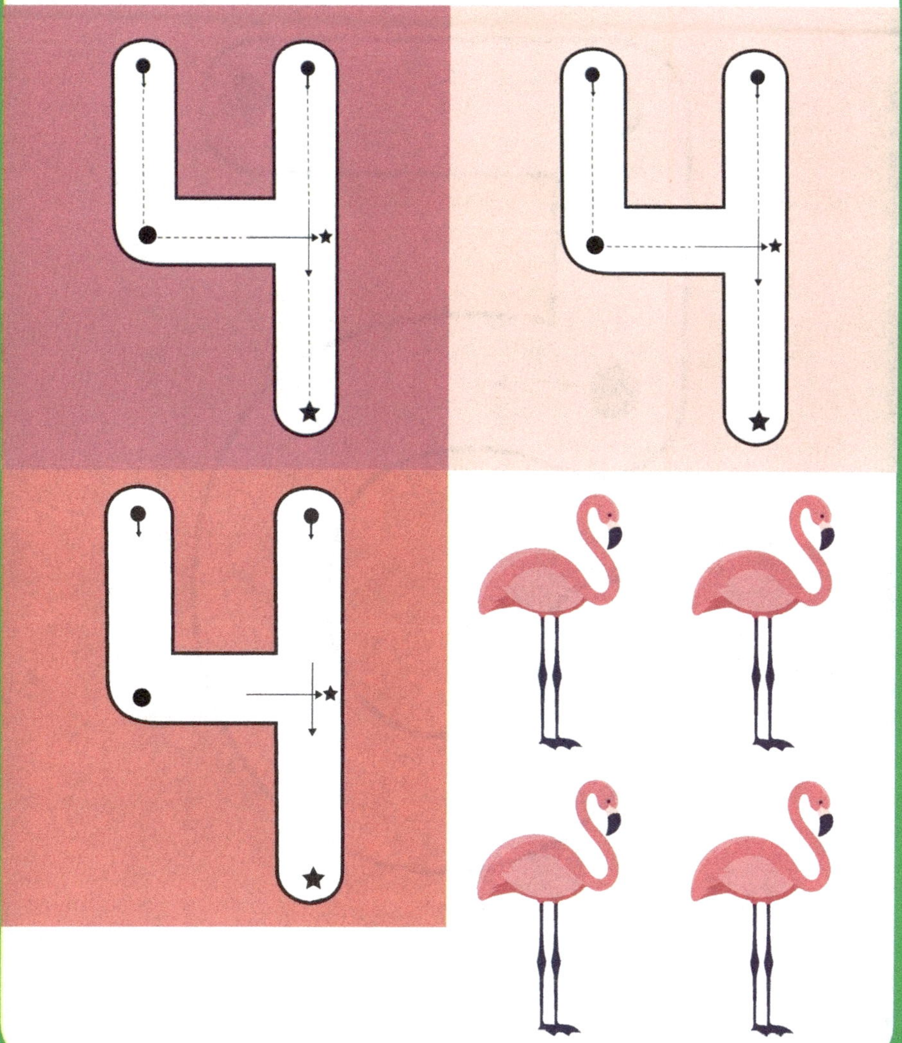

LEMONLearn

Leaf Math
Booklet 19
Tracing 1
(up to 10)

Trace the number.

FIVE

LEMONLearn

Trace the numbers.

FIVE

Trace the number.

SIX

Trace the numbers.

SIX

Trace the number.

SEVEN

Trace the numbers.

SEVEN

Trace the number.

EIGHT

Trace the numbers.

EIGHT

Trace the number.

NINE

Trace the numbers.

NINE

LEMONLearn

Leaf Math
Booklet 19
Tracing 1
(up to 10)

Trace the number.

TEN

190b

LEMONLearn

Leaf Math
Booklet 19
Tracing 1
(up to 10)

Trace the numbers.

TEN

COLOR BY NUMBERS

- 1 YELLOW
- 2 LIGHT GREEN
- 3 LIGHT PINK
- 4 LIGHT BLUE
- 5 ORANGE

- **6 DARK PINK**
- **7 DARK GREEN**
- **8 RED**
- **9 DARK BLUE**
- **10 PURPLE**

Booklet 20

Tracing up to 10

1. Preview the pages of Booklet 20 with your student before starting.
2. Help your student write their name, date, and start / finish time.
3. Work together with your student for 15-30 minutes daily.

Name: ______________________

Date: ______________________

Time: ____ : ____ **to** ____ : ____

LEMONLearn

Leaf Math
Booklet 20
Tracing 2
(up to 10)

Trace the number.

ONE

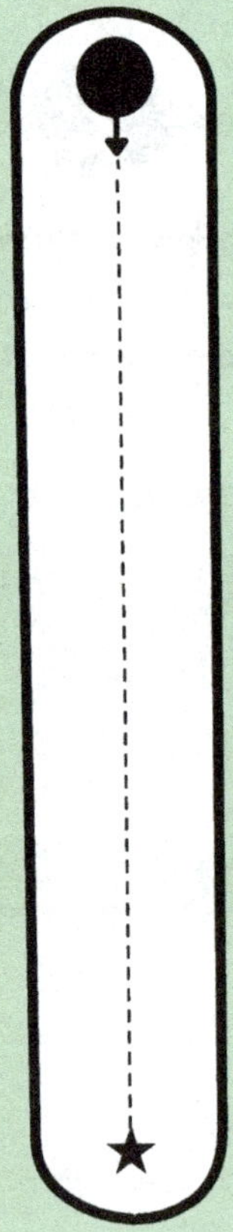

Trace the numbers.

ONE

LEMONLearn

Trace the number.

TWO

LEMONLearn

Leaf Math
Booklet 20
Tracing 2
(up to 10)

Trace the numbers.

TWO

LEMONLearn

Trace the number.

THREE

Trace the numbers.

THREE

LEMONLearn

Trace the number.

FOUR

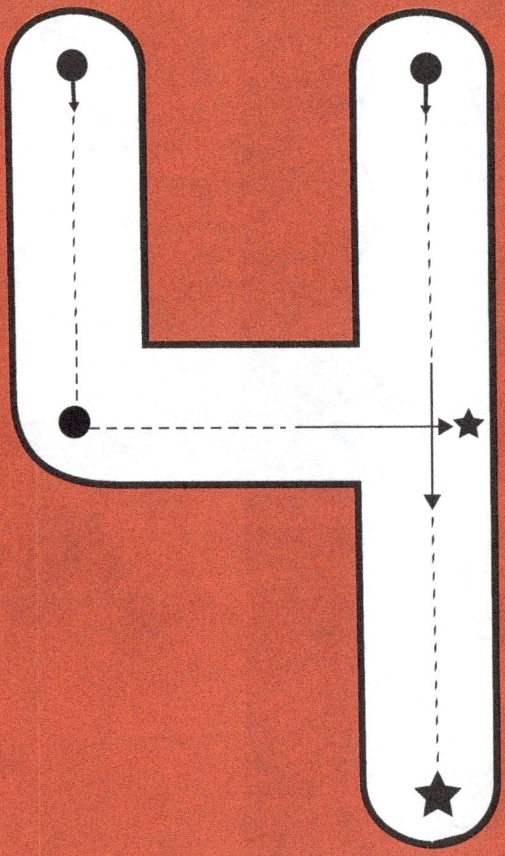

LEMONLearn

Trace the numbers.

FOUR

Trace the number.

FIVE

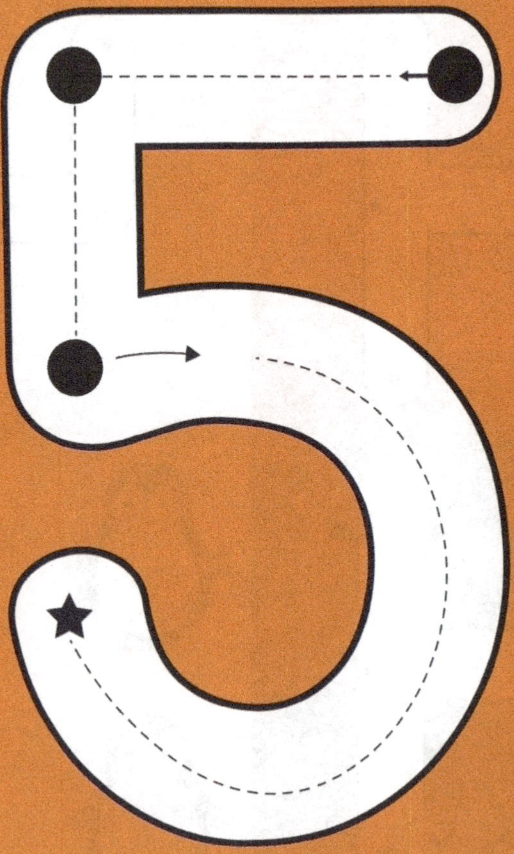

Trace the numbers.

FIVE

Trace the number.

SIX

LEMONLearn

Leaf Math
Booklet 20
Tracing 2
(up to 10)

Trace the numbers.

SIX

Trace the number.

SEVEN

Trace the numbers.

SEVEN

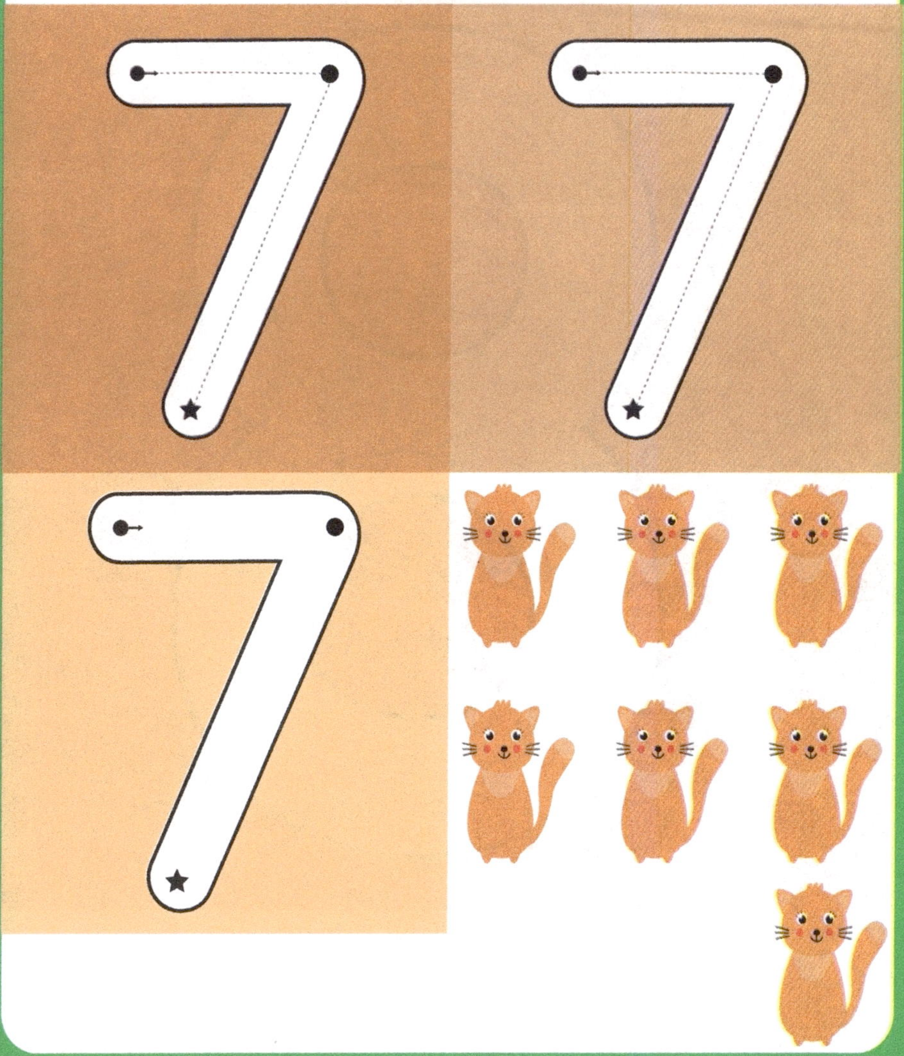

Trace the number.

EIGHT

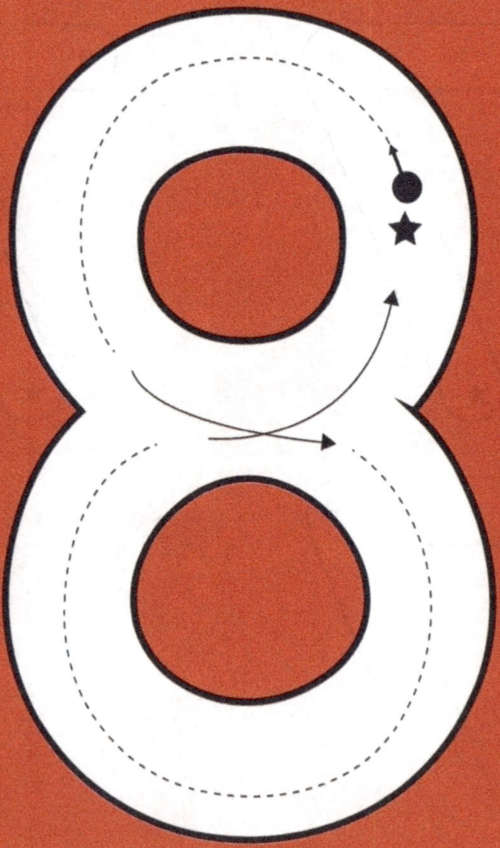

LEMONLearn

Trace the numbers.

EIGHT

Trace the number.

NINE

LEMONLearn

Leaf Math
Booklet 20
Tracing 2
(up to 10)

Trace the numbers.

NINE

Trace the number.

TEN

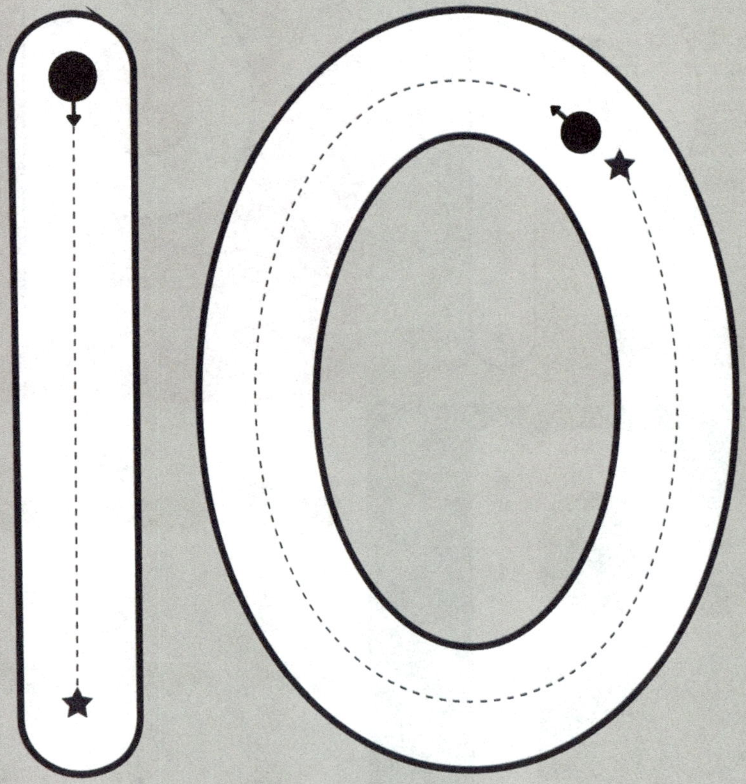

Trace the numbers.

TEN

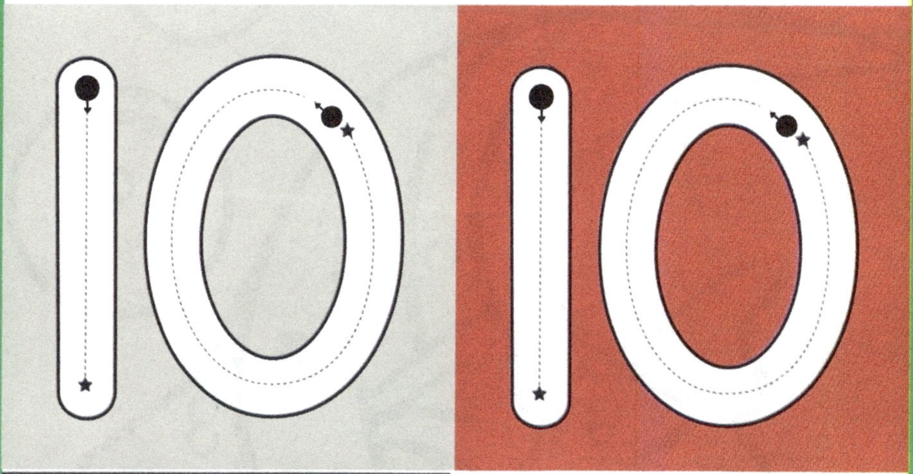

COLOR BY NUMBERS

- 1 YELLOW
- 2 LIGHT GREEN
- 3 LIGHT PINK
- 4 LIGHT BLUE
- 5 ORANGE

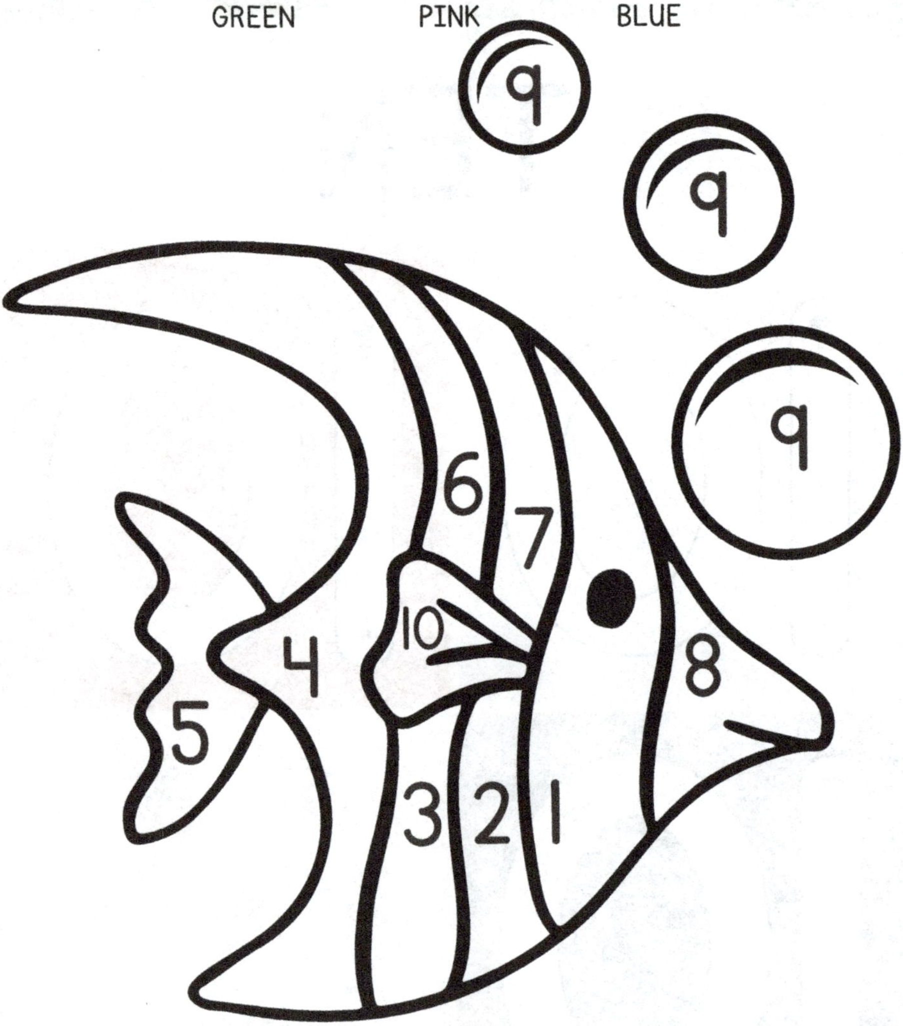

- 6 DARK PINK
- 7 DARK GREEN
- 8 RED
- 9 DARK BLUE
- 10 PURPLE

Booklet 21

Tracing up to 10

1. Preview the pages of Booklet 21 with your student before starting.
2. Help your student write their name, date, and start / finish time.
3. Work together with your student for 15-30 minutes daily.

Name: ______________________

Date: ______________________

Time: ____ : ____ **to** ____ : ____

Trace the number.

ONE

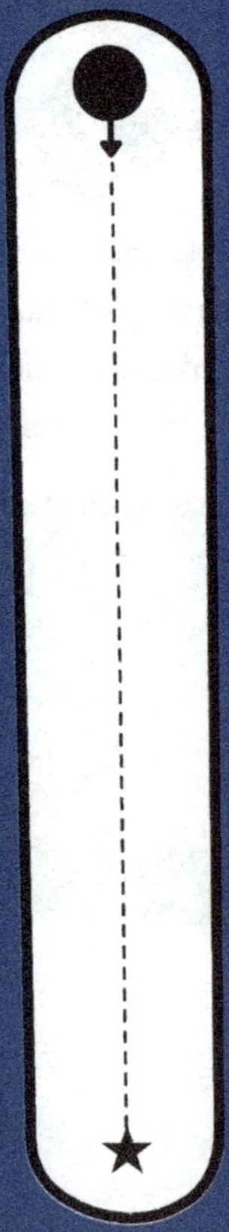

LEMONLearn

Trace the numbers.

ONE

LEMONLearn

Trace the number.

TWO

LEMONLearn

Leaf Math
Booklet 21
Tracing 3
(up to 10)

Trace the numbers.

TWO

LEMONLearn

Trace the number.

THREE

203b

LEMONLearn

Leaf Math
Booklet 21
Tracing 3
(up to 10)

Trace the numbers.

THREE

Trace the number.

FOUR

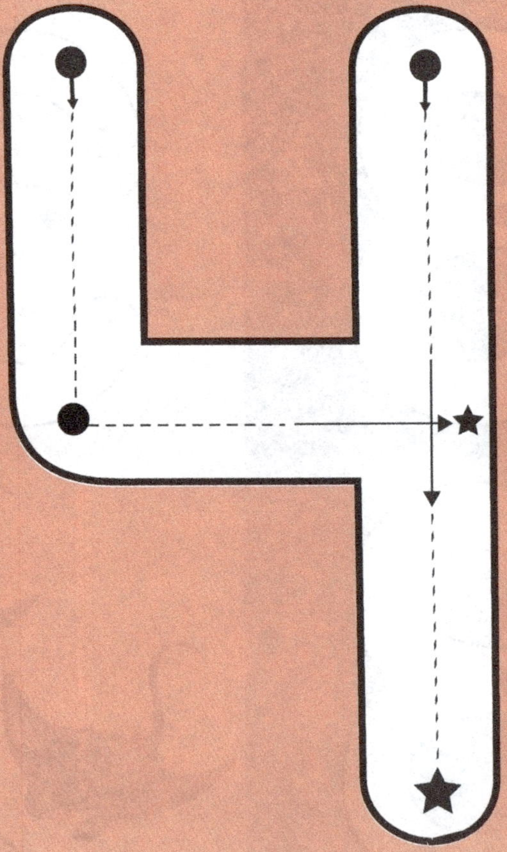

Trace the numbers.

FOUR

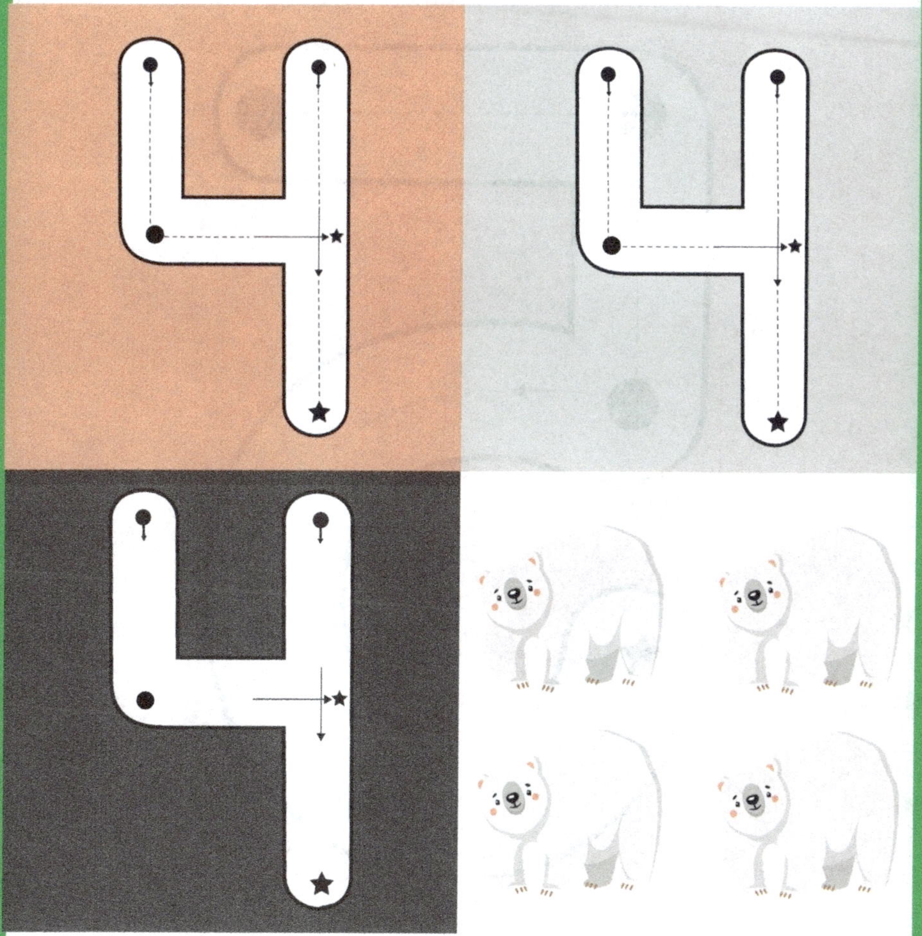

205a

LEMONLearn

Leaf Math
Booklet 21
Tracing 3
(up to 10)

Trace the number.

FIVE

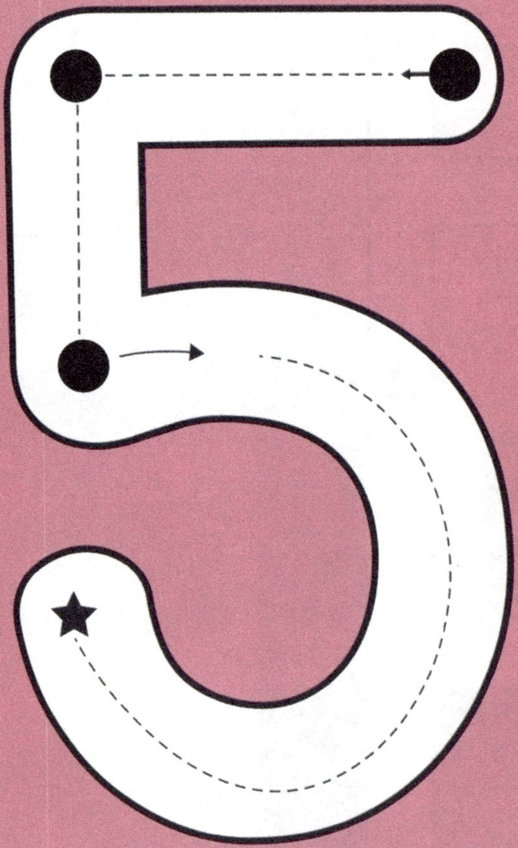

Trace the numbers.

FIVE

Trace the number.

SIX

LEMONLearn

Trace the numbers.

SIX

LEMONLearn

Trace the number.

SEVEN

207b

LEMONLearn

Leaf Math

Booklet 21

Tracing 3

(up to 10)

Trace the numbers.

SEVEN

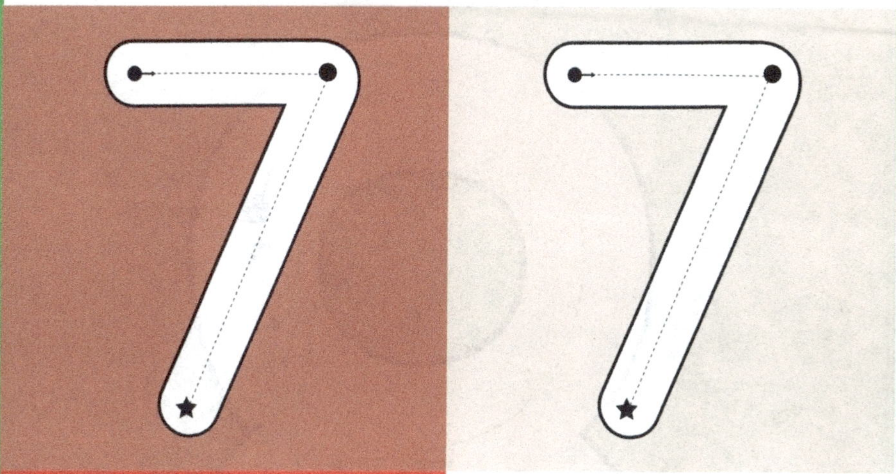

Trace the number.

EIGHT

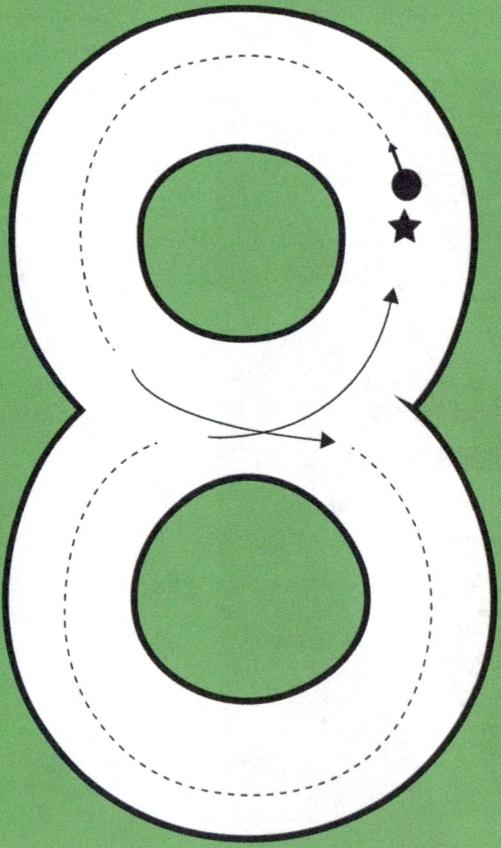

Trace the numbers.

EIGHT

LEMONLearn

Trace the number.

NINE

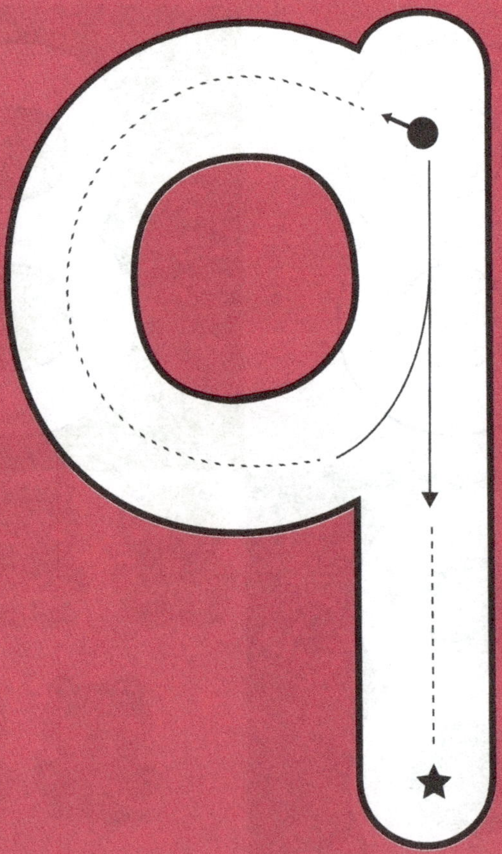

LEMONLearn

Leaf Math
Booklet 21
Tracing 3
(up to 10)

Trace the numbers.

NINE

Trace the number.

TEN

LEMONLearn

Trace the numbers.

TEN

COLOR BY NUMBERS

1 YELLOW

2 LIGHT GREEN

3 LIGHT PINK

4 LIGHT BLUE

5 ORANGE

6 **DARK PINK**

7 **DARK GREEN**

8 **RED**

9 **DARK BLUE**

10 **PURPLE**

1 2 3 4 5 6 7 8 9 10

Congratulations

LEMON LEARNER

NAME

you DID IT.

PROUDLY PRESENTED BY

LEMON Learn

YOUR NEXT BOOK

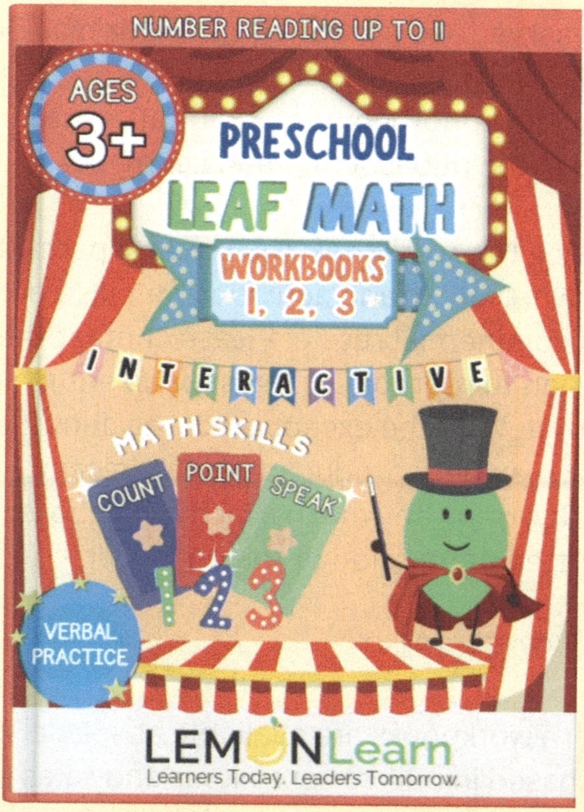

Continue your child's learning with the next book in the LEAF Math Enrichment Program.

Preschool LEAF Math Workbooks 1, 2, 3

- Designed for early learners ages 3 and up.
- This book introduces number 11 as the next step after mastering numbers up to 10, helping students transition smoothly while building confident number progression.
- A color by numbers pages at the at the end of each section adds hands-on learning and fun.
- An achievement award at the end of the book celebrates your child's hard work and learning success.

ISBN 978-1-971510-91-0

Lemon Learn workbooks are available at special discounts when purchased in bulk for premiums and sales promotions as well as for fundraising or educational use. Special editions or book excerpts can also be created to specification. For details please contact lemonlearn123@gmail.com

Parents and educators, visit us on the Web for a variety of teaching tools!

lemonlearn.com

LEMONLearn
Learners Today. Leaders Tomorrow.

Made in the USA
Monee, IL
03 November 2026

67380761R00046